UNDERSTANDING THE LIMITS OF ARTIFICIAL INTELLIGENCE FOR WARFIGHTERS

VOLUME 1_ SUMMARY

LANCE MENTHE

LI ANG ZHANG

EDWARD GEIST

JOSHUA STEIER

AARON B. FRANK

ERIK VAN HEGEWALD

GARY J. BRIGGS

KELLER SCHOLL

YUSUF ASHPARI

ANTHONY JACQUES

PREPARED FOR THE DEPARTMENT OF THE AIR FORCE
APPROVED FOR PUBLIC RELEASE; DISTRIBUTION IS UNLIMITED.

 PROJECT AIR FORCE

01

For more information on this publication, visit **www.rand.org/t/RRA1722-1**.

About RAND

RAND is a research organization that develops solutions to public policy challenges to help make communities throughout the world safer and more secure, healthier and more prosperous. RAND is nonprofit, nonpartisan, and committed to the public interest. To learn more about RAND, visit www.rand.org.

Research Integrity

Our mission to help improve policy and decisionmaking through research and analysis is enabled through our core values of quality and objectivity and our unwavering commitment to the highest level of integrity and ethical behavior. To help ensure our research and analysis are rigorous, objective, and nonpartisan, we subject our research publications to a robust and exacting quality-assurance process; avoid both the appearance and reality of financial and other conflicts of interest through staff training, project screening, and a policy of mandatory disclosure; and pursue transparency in our research engagements through our commitment to the open publication of our research findings and recommendations, disclosure of the source of funding of published research, and policies to ensure intellectual independence. For more information, visit www.rand.org/about/research-integrity.

RAND's publications do not necessarily reflect the opinions of its research clients and sponsors.

Library of Congress Cataloging-in-Publication Data is available for this publication.

ISBN: 978-1-9774-1278-2

Cover: Master Sgt. Matthew Plew/U.S. Air Force and your123/Adobe Stock.

Limited Print and Electronic Distribution Rights

About This Report

This is the first report in a five-volume series addressing how artificial intelligence (AI) could be employed to assist warfighters in four distinct areas: cybersecurity, predictive maintenance, wargames, and mission planning. These areas were chosen to reflect the wide variety of potential uses and to highlight different kinds of limits to AI application. Each use case is presented in a separate volume, as it will be of interest to a different community.

This first volume summarizes the findings and recommendations from all use cases. It is aimed at policymakers, acquisition professionals, and those with a general interest in the application of AI to warfighting. The following other volumes provide detailed analysis of the individual use cases:

- Joshua Steier, Erik Van Hegewald, Anthony Jacques, Gavin S. Hartnett, and Lance Menthe, *Understanding the Limits of Artificial Intelligence for Warfighters:* Vol. 2, *Distributional Shift in Cybersecurity Datasets*, RR-A1722-2, 2024
- Li Ang Zhang, Yusuf Ashpari, and Anthony Jacques, *Understanding the Limits of Artificial Intelligence for Warfighters:* Vol. 3, *Predictive Maintenance*, RR-A1722-3, 2024
- Edward Geist, Aaron B. Frank, and Lance Menthe, *Understanding the Limits of Artificial Intelligence for Warfighters:* Vol. 4, *Wargames*, RR-A1722-4, 2024
- Keller Scholl, Gary J. Briggs, Li Ang Zhang, and John L. Salmon, *Understanding the Limits of Artificial Intelligence for Warfighters:* Vol. 5, *Mission Planning*, RR-A1722-5, 2024.

The research reported here was commissioned by Air Force Materiel Command, Strategic Plans, Programs, Requirements and Assessments (AFMC/A5/8/9) and conducted within the Force Modernization and Employment Program of RAND Project AIR FORCE as part of a fiscal year 2022 project, "Understanding the Bounds of Artificial Intelligence in Warfare Applications."

RAND Project AIR FORCE

RAND Project AIR FORCE (PAF), a division of the RAND Corporation, is the Department of the Air Force's (DAF's) federally funded research and development center for studies and analyses, supporting both the United States Air Force and the United States Space Force. PAF provides the DAF with independent analyses of policy alternatives affecting the development, employment, combat readiness, and support of current and future air, space, and cyber forces. Research is conducted in four programs: Strategy and Doctrine; Force Modernization and Employment; Resource Management; and Workforce, Development, and Health. The research reported here was prepared under contract FA7014-22-D-0001.

Additional information about PAF is available on our website:
www.rand.org/paf/

This report documents work originally shared with the DAF on September 23, 2022. The draft report, dated September 2022, was reviewed by formal peer reviewers and DAF subject-matter experts.

Acknowledgments

We thank our sponsor contact, Kathryn Sowers, and our action officers, Julia Phillips and Gregory Cazzell, for their guidance in choosing the use cases, for their thoughtfulness in scoping the research questions, and for working diligently with us to obtain the data necessary to conduct the many machine-learning experiments described in this series of reports. Thanks as well to the following individuals: Jeremy Brogdon for his assistance in obtaining and understanding predictive maintenance issues; Richard Moore for sharing his immense expertise in these matters; R. Scott Erwin and Jean-Charles Ledé for graciously connecting us with many AI development efforts across the Air Force Research Laboratory; Lee Seversky for sharing his expertise on mission planning; David A. Kapp for discussing distributional shift; Lt Col Kari K. Mott, who discussed automation in Air Operation Centers today; and Lt Col Nicholas J. Harris, who spoke with us across a dozen time zones to explain the Master Air Attack Planning process.

We are also grateful to many current and former RAND colleagues, including Caolionn O'Connell, Sherrill Lingel, Osonde Osoba, and Chris Pernin for helping us shape the research agenda. Thanks to John Salmon for sharing his insights and to Matthew Walsh, Marjory S. Blumenthal, and Jair Aguirre for their thoughtful reviews of this document. Great appreciation to John Drew and James Leftwich for offering their invaluable expertise on predictive maintenance issues, to Ellie Bartels and Paul K. Davis for sharing their expertise on wargaming, and to Angela Gruber of the Prince Rupert Port Authority for discussing her novel application of AI to the problem of identifying conjoined moving objects. We could not have written these reports without their help; any errors that remain are ours alone.

Summary

Issue

The U.S. Air Force has become increasingly interested in the potential for artificial intelligence (AI) to enhance different aspects of warfighting. For this project, the Air Force asked the RAND Corporation to consider instead what AI *cannot* do in order to understand the limits of AI for warfighting applications.

Approach

Rather than attempt to determine the limits of AI in general, we selected and investigated four specific warfighting applications as potential use cases: *cybersecurity, predictive maintenance, wargames,* and *mission planning.* These applications were chosen to represent a variety of possible uses while highlighting different constraints. We tailored the research approach to each use case. In the three cases for which we believed we could obtain sufficient data, we performed AI experiments; in the remaining case, wargames, we looked broadly at how AI could or could not be applied. The details for each use case are presented in separate volumes.

Key Findings

Two common themes emerge from the use cases: (1) data to train and test AI systems must be current, accessible, and of high quality; and (2) the limitations of AI algorithms can significantly restrict their utility. Table S.1 summarizes the major findings for each use case.

Table S.1. Summary of Findings

Use Case	Data Limitations	Algorithm Limitations
Cybersecurity	• **To recognize adaptive threats, data must be recent.** Distributional shift—the growing gap between real-world experience and initial training data—degrades model performance, and it cannot be avoided in malware or network intrusion, especially for high-dimensional data.	• **AI classification algorithms cannot be relied on to learn what they are not taught.** AI did not anticipate or recognize new kinds of malware or network intrusion.
Predictive maintenance	• **Data must be accessible and well-conditioned.** Relevant logistics data are maintained in multiple databases and are often ill-conditioned. Without an automated data pipeline, sufficient data cannot be captured to enable AI. • **Peacetime data cannot be substituted for wartime data.** AI cannot make up for a scarcity of appropriate data.	• **AI can estimate complex functions well, but this comes at a loss of generality.** AI can estimate part failure rates far better than a universal probability distribution, but it must be trained on each part separately.
Wargames	• **Digitization must precede AI development.** Most wargames are not conducted in a digital environment and do not generate electronic data. Digitization is a precursor to an AI data pipeline. • **New kinds of data are needed.** To enable AI, human-computer interaction (HCI) technology is needed to capture aspects of wargaming that are not captured today.	• **AI is far from achieving human-level intelligence.** Therefore, it cannot stand in for humans, nor can it apply human judgments. AI is therefore only likely applicable to certain stages of wargames conducted for certain purposes.
Mission planning	• **To counter adaptive threats, data must be recent.** Models must be refreshed with updated conditions to survive against dynamic threats.	• **AI is tactically brilliant but strategically naive.** It tends to win by getting within the opponent's observe, orient, decide, act loop rather than by coming up with a clever grand strategy. • **AI is less accurate than traditional optimization methods.** But its solutions can be more robust, and it can reach them faster.

Recommendations

In the recommendations across all use cases, two themes emerged: the need to conduct tests and experiments and the need to develop better infrastructure to support future AI development. Table S.2 summarizes the recommendations, specific to each use case.

Table S.2. Summary of Recommendations

Use Case	AI Tests and Experiments	Supporting Infrastructure
Cybersecurity	• Perform dataset segmentation tests to determine the significance of distributional shift for AI systems and to determine an approximate decay rate and AI shelf life.	• Not applicable.
Predictive maintenance	• Experiment with AI to improve demand forecasting for readiness spares packages (RSPs) and extend the proof-of-concept models to all aircraft. This will likely have to be done on a part-by-part, platform-by-platform basis. • Consider AI to solve the larger operations research problem of selecting which parts to send where.	• Build a data operations pipeline to conduct a retrospective analysis of aircraft maintenance and RSP efficiently for multiple parts and platforms.
Wargames	• Concentrate resources for developing AI applications for wargames on the most promising areas: those that investigate alternative conditions or that are used for evaluation with well-defined criteria; those that already incorporate digital infrastructure, including HCI technologies; and those that are regularly repeated.	• Increase the use of digital gaming infrastructure and HCI technologies, especially in games designed for systems exploration and innovation, to gather data to support AI development. • Employ AI capabilities to support future wargaming efforts more generally.
Mission planning	• Consider how AI could power a fast-reaction policy for drones facing unexpected conditions.	• Invest in developing tools to apply reinforcement learning to existing mission planning models and in simulations, such as the Advanced Framework for Simulation, Integration, and Modeling (AFSIM).

Contents

About This Report..iii

Summary..v

Figures and Tables..ix

CHAPTER 1...1

Overview..1

    Introduction..1

    What Do We Mean by Artificial Intelligence?..2

    Selecting the Use Cases...3

    How to Read These Reports..5

CHAPTER 2...6

Approach, Findings, and Recommendations...6

    Cybersecurity...6

    Predictive Maintenance...8

    Wargames..12

    Mission Planning..15

    Conclusion..17

Abbreviations...20

References...21

Figures and Tables

Figures

Figure 2.1. Comparison of Actual Part Failure Rates for the A-10C from 2007 to 2022 with Predictions by the Aircraft Sustainability Model and Artificial Intelligence Model 10

Figure 2.2. Technical Feasibility and Cost-Effectiveness of Developing Artificial Intelligence for Wargames... 14

Figure 2.3. Screenshots of the RAND Target Acquisition Model and Advanced Framework for Simulation, Integration, and Modeling 16

Tables

Table S.1. Summary of Findings vi

Table S.2. Summary of Recommendations vii

Table 1.1. Selection Criteria 4

Table 1.2. Artificial Intelligence Use Selection 4

Table 2.1. Taxonomy of Wargames by Purpose 13

Table 2.2. Summary of Findings 18

Table 2.3. Summary of Recommendations 19

Chapter 1

Overview

Introduction

The Department of the Air Force (DAF) has become increasingly interested in the potential for artificial intelligence (AI) to enhance different aspects of warfighting. As Secretary of Defense Lloyd Austin said in 2021, "AI is central to our innovation agenda, helping us to compute faster, share better . . . and make decisions faster and more rigorously."[1] Over the past five years, RAND Corporation researchers have investigated how AI might be used to improve command and control, intelligence analysis, operational assessment, human resources management, and many other applications.[2] For this project, the U.S. Air Force (USAF) asked RAND to consider instead what AI *cannot* do in order to understand the limits of AI for warfighting applications.

Of course, determining the limits of AI for all possible warfighting applications would be too large of a subject to address in any one report. Rather than attempting to determine the limits of AI in general, we selected and investigated four specific warfighting applications as potential use cases: *cybersecurity, predictive maintenance, wargames,* and *mission planning.* These applications were chosen in consultation with our sponsor, Air Force Materiel Command, Strategic, Plans Programs, Requirements, and Analyses (AFMC/A5/8/9), to represent a variety of possible AI uses in the DAF while highlighting different constraints on AI use.

In this report, we describe various inherent limits to using AI to help solve different problems, including data and algorithmic issues. However, there is an important category of limitations that we do not consider: the potential for *adversarial attacks* to disrupt AI systems, such as generative

---

[1] Lloyd J. Austin III, "Secretary of Defense Austin Remarks at the Global Emerging Technology Summit of the National Security Commission on Artificial Intelligence (as Delivered)," transcript, U.S. Department of Defense, July 13, 2021.

[2] Matthew Walsh, Lance Menthe, Edward Geist, Eric Hastings, Joshua Kerrigan, Jasmin Léveillé, Joshua Margolis, Nicholas Martin, and Brian P. Donnelly, *Exploring the Feasibility and Utility of Machine Learning-Assisted Command and Control: Vol. 1, Findings and Recommendations,* RAND Corporation, RR-A263-1, 2021a; Sherrill Lingel, Jeff Hagen, Eric Hastings, Mary Lee, Matthew Sargent, Matthew Walsh, Li Ang Zhang, and David Blancett, *Joint All-Domain Command and Control for Modern Warfare: An Analytic Framework for Identifying and Developing Artificial Intelligence Applications,* RAND Corporation, RR-4408/1-AF, 2020; Daniel Ish, Jared Ettinger, and Christopher Ferris, *Evaluating the Effectiveness of Artificial Intelligence Systems in Intelligence Analysis,* RAND Corporation, RR-A464-1, 2021; Lance Menthe, Dahlia Anne Goldfeld, Abbie Tingstad, Sherrill Lingel, Edward Geist, Donald Brunk, Amanda Wicker, Sarah Lovell, Balys Gintautas, Anne Stickells, and Amado Cordova, *Technology Innovation and the Future of Air Force Intelligence Analysis: Vol. 1, Findings and Recommendations,* RAND Corporation, RR-A341-1, 2021a; Daniel Egel, Ryan Andrew Brown, Linda Robinson, Mary Kate Adgie, Jasmin Léveillé, and Luke J. Matthews, *Leveraging Machine Learning for Operation Assessment,* RAND Corporation, RR-4196-A, 2022; David Schulker, Nelson Lim, Luke J. Matthews, Geoffrey E. Grimm, Anthony Lawrence, and Perry Shameem Firoz, *Can Artificial Intelligence Help Improve Air Force Talent Management? An Exploratory Application,* RAND Corporation, RR-A812-1, 2021; Sean Robson, Maria C. Lytell, Matthew Walsh, Kimberly Curry Hall, Kirsten M. Keller, Vikram Kilambi, Joshua Snoke, Jonathan W. Welburn, Patrick S. Roberts, Owen Hall, and Louis T. Mariano, *U.S. Air Force Enlisted Classification and Reclassification: Potential Improvements Using Machine Learning and Optimization Models,* RAND Corporation, RR-A284-1, 2022.

adversarial networks, data poisoning, so-called Trojan algorithms, and others. Those limitations are discussed in other RAND reports.[3]

What Do We Mean by Artificial Intelligence?

Updating Marvin Minksy's original 1968 definition, we echo prior RAND reports that define AI broadly as "the use of computers to carry out tasks that previously required human intelligence."[4] For the purposes of discussing warfighting applications, we also follow prior RAND practice and consider AI to consist of these six specific capabilities:

- *computer vision*, the detection and classification of objects in visual media[5]
- *natural language processing*, the recognition and translation of speech and text
- *planning*, the use of models to find actions that lead to goals
- *prediction and classification*, the categorization of current and future data based on prior data
- *generative learning*, the synthesis of language, images, and other media
- *expert systems*, rules-based models constructed to reflect expert knowledge and general heuristics.[6]

In this series of reports, we look almost exclusively at the potential to use current machine-learning (ML) methods—primarily, neural networks—to achieve these AI capabilities. ML and AI are not synonymous, although they are often treated as such because the rapid progress in AI capabilities over the past decade has largely been the result of the revolution in deep learning, a subfield of ML that has rendered obsolete prior approaches to computer vision and natural language

---

[3] For example, Andrew J. Lohn, Jair Aguirre, Mark Ashby, Benjamin Boudreaux, Johnathan Fujiwara, Gavin S. Hartnett, Daniel Ish, John Speed Meyers, Caolionn O'Connell, and Li Ang Zhang, *Attacking Machine Learning in War*, RAND Corporation, RR-4386-AF, 2020, Not available to the general public; and Li Ang Zhang, Gavin S. Hartnett, Jair Aguirre, Andrew J. Lohn, Inez Khan, Marissa Herron, and Caolionn O'Connell, *Operational Feasibility of Adversarial Attacks Against Artificial Intelligence*, RAND Corporation, RR-A866-1, 2022.

[4] Minsky originally defined AI as "the science of making machines do things that would require intelligence if done by men" (Marvin Minsky, ed., *Semantic Information Processing*, MIT Press, 1968, p. v); Lance Menthe, Dahlia Anne Goldfeld, Abbie Tingstad, Sherrill Lingel, Edward Geist, Donald Brunk, Amanda Wicker, Sarah Lovell, Balys Gintautas, Anne Stickells, and Amado Cordova, *Technology Innovation and the Future of Air Force Intelligence Analysis: Vol. 2, Technical Analysis and Supporting Material*, RAND Corporation, RR-A341-2, 2021b, p. 46.

[5] Although computer vision is technically a subset of prediction and classification, its techniques are sufficiently specialized and subject to unique vulnerabilities that we treat it as its own category for these purposes.

[6] These capabilities were derived from a standard AI reference, Stuart Russell and Peter Norvig, *Artificial Intelligence: A Modern Approach*, 4th ed., Pearson, 2021. See also Li Ang Zhang, Lance Menthe, Ian Fleischmann, Sale Lilly, Joshua Kerrigan, Michael J. Gaines, and Gregory A. Schumacher, *Incorporating Artificial Intelligence into Army Intelligence Processes*, RAND Corporation, RR-A729-1, 2021, Not available to the general public.

processing.[7] *Deep learning* is a loose term referring to neural networks that have many "hidden" interconnected layers of "neurons" between the input and output streams.[8]

Modern ML is also commonly spoken of in terms of three subcategories: supervised learning, in which the AI system is trained using labeled data; unsupervised learning, in which the AI system seeks to uncover hidden structures in unlabeled data; and reinforcement learning, in which the AI system is rewarded for maximizing an expected utility function through its interactions with its environment.[9] We consider all three in this report but look mostly at supervised and reinforcement learning, which might not require more data than unsupervised learning but which do usually require the data to be processed or labeled.

We do not look at the potential for artificial general intelligence or any other groundbreaking new methods in ML. Although there is no standard definition for artificial general intelligence, it is more or less the hypothetical ability to achieve cognition that is "human-like" or "general" in the sense that it is not tied to specific tasks. There is no consensus as to what technologies could achieve this ability, and many experts believe that fundamental breakthroughs are needed first.[10] As one recent observer put it,

> Machines may someday be as smart as people, and perhaps even smarter, but the game is far from over. There is still an *immense* [emphasis added] amount of work to be done in making machines that truly can comprehend and reason about the world around them. . . . This is why basic research remains crucial.[11]

We are mindful, however, that it is the nature of breakthrough research to defy prediction. When we speak of the limits of AI for warfighting applications, therefore, we wish to be explicit that we refer only to the limits of the field as we understand it today. Describing the limits of AI today might be similar to describing the limits of electricity in 1950, when nuclear-powered electric cars seemed just as plausible as flying cars. Still, we expect these predictions to be useful for planning over a 20-year time frame.

Selecting the Use Cases

The first task in this project was to select potential AI use cases that were of interest to the DAF and were likely to highlight different types of limits to AI use. We first proposed the five selection criteria shown in Table 1.1.

---

[7] Prior approaches to computer vision relied largely on feature extraction methods, such as edge detection and corner detection. See Niall O'Mahony, Sean Campbell, Anderson Carvalho, Suman Harapanahalli, Gustavo Velasco Hernández, Lenka Krpalkova, Daniel Riordan, and Joseph Walsh, "Deep Learning vs. Traditional Computer Vision," in Kohei Arai and Supriya Kapoor, eds, *Advances in Computer Vision: Proceedings of the 2019 Computer Vision Conference CVC*, Vol. 1, Springer, 2019; Kohei Arai and Supriya Kapoor, eds, *Advances in Computer Vision: Proceedings of the 2019 Computer Vision Conference CVC*, Vol. 1, Springer, 2019.

[8] M. L. Jordan and T. M. Mitchell, "Machine Learning: Trends, Perspectives, and Prospects," *Science*, Vol. 349, No. 6245, July 2015.

[9] Christopher M. Bishop, *Pattern Recognition and Machine Learning*, Springer, 2006, p. 3.

[10] Menthe et al., 2021b.

[11] Gary Marcus, "Artificial General Intelligence Is Not as Imminent as You Might Think," *Scientific American*, July 1, 2022.

Table 1.1. Selection Criteria

Criterion	Description
DAF importance	How important to the DAF is the capability gap or process that this AI application addresses?
DAF impact	How much could an AI application theoretically improve efficiency, effectiveness, use of human capital, and/or agility in this process?
AI success	How much success has AI been shown to have in achieving these kinds of improvements in related areas?
AI limit relevance	How likely is it that this use case will highlight an important limitation or boundary of AI application for the warfighter?
Research feasibility	How feasible is it for RAND to conduct ML experiments in this area?

We then proposed 14 research areas, varying from intelligence, surveillance, and reconnaissance (ISR) applications, to command and control (C2) and operations, to enterprise uses, such as logistics and personnel.[12] These were a combination of areas in which AI work was already active in the DAF or at RAND or in which the subject-matter experts (SMEs) we consulted felt there was untapped potential. Table 1.2 shows the potential research areas by category, along with our rough rankings according to the five selection criteria.

Table 1.2. Artificial Intelligence Use Selection

Category	Use Case	Selection Criteria				
		Importance to DAF	Potential Impact of AI	Likelihood of AI Success	AI Limit Relevance	Feasibility of Research
ISR	Tactical Multi-language Transcription	M	H	H	M	H
	Full Motion Video Exploitation	H	H	H	H	M
	OSINT Data for Battle Damage Assessment	M	H	H	M	H
	Automated Identification of Friend-or-Foe	H	M	M	H	M
	Tracking Adversary Disposition of Forces	H	M	H	H	H
	Dynamically Optimized Sensing Grid	M	H	M	L	M
C2/Operations	Wargames	M	M	M	H	H
	Mission Planning	H	M	H	M	H
	Defensive Counterspace	M	H	L	H	H
	OSINT-informed PSYOPS	L	M	H	M	M
Enterprise	Cybersecurity	M	H	M	H	M
	Predictive Maintenance	M	H	H	M	M
	Promotion Selection Criteria Analysis	M	M	M	H	M
	Repurpose Legacy PowerPoint Data	L	H	H	L	H

H High   M Medium   ▉ Low

NOTE: OSINT = open source intelligence; PSYOPS = psychological operations.

---

[12] We initially proposed 13 research areas. The wargames use case was suggested by our sponsor in one of the initial meetings, during which we talked through the potential research areas.

While we hoped that the selection criteria would yield obvious candidates, a quick examination of Table 1.2 shows that the results are mixed and there are no obvious winners. AI is broadly applicable, and many areas of research are interesting, but every area has some issues as well. In the end, we decided (in consultation with our sponsor) to rule out the entire ISR category because RAND researchers have already done considerable work in this area and because there were ongoing studies involving the disposition of forces and the sensing grid.[13] Nevertheless, this is an important area for further research. We also ruled out any area with a "low" ranking in any category. The last area we omitted was promotion criteria selection analysis because it had the fewest "high" ratings and because RAND researchers had recently investigated this area.[14]

That left us with the four use cases: cybersecurity, predictive maintenance, wargames, and mission planning. These cases covered a variety of options and posed different research challenges.

How to Read These Reports

Chapter 2 presents a summary of the research approach, findings, and recommendations from each use case. The subsequent four volumes in the series discuss each use case separately and in more detail.[15] We offer these details in distinct volumes because we expect the use cases might be of interest to different warfighting communities.

---

[13] For example, see Menthe et al., 2021a; and Menthe et al., 2021b.

[14] Schulker et al., 2021.

[15] See Joshua Steier, Erik Van Hegewald, Anthony Jacques, Gavin S. Hartnett, and Lance Menthe, *Understanding the Limits of Artificial Intelligence for Warfighters: Vol. 2, Distributional Shift in Cybersecurity Datasets*, RAND Corporation, RR-A1722-2, 2024; Li Ang Zhang, Yusuf Ashpari, and Anthony Jacques, *Understanding the Limits of Artificial Intelligence for Warfighters: Vol. 3, Predictive Maintenance*, RAND Corporation, RR-A1722-3, 2024; Edward Geist, Aaron B. Frank, and Lance Menthe, *Understanding the Limits of Artificial Intelligence for Warfighters: Vol. 4, Wargames*, RAND Corporation, RR-A1722-4, 2024; and Keller Scholl, Gary J. Briggs, Li Ang Zhang, and John L. Salmon, *Understanding the Limits of Artificial Intelligence for Warfighters: Vol. 5, Mission Planning*, RAND Corporation, RR-A1722-5, 2024.

Chapter 2

Approach, Findings, and Recommendations

We tailored our research approach to each use case. In the three cases for which large datasets are publicly available, we drilled down to particular tasks so we could perform technical experiments on the data. In the remaining case, wargames, we developed a framework for analysis and looked broadly at how AI could or could not be applied in the future. In this chapter, we describe each use case separately and conclude with a summary of the common themes.

Cybersecurity

The research objective for this use case was to consider how the potential for distributional shift could limit AI effectiveness for cybersecurity applications.

Distributional shift occurs when the data that an AI system encounters in the field diverge over time from the data on which the AI system was trained and tested. Depending on the precise nature of the shift, this can significantly degrade AI system performance over time. We focused on distributional shift because recent research indicated that distributional shift was likely to be a significant factor for AI cybersecurity applications. As one research group explained, "In the real network security, hackers often change different ways to attack the network, in such cases, there could be multiple variables changes might have happened during the period."[16]

Distributional shift is particularly challenging in such cases where the shift is unpredictable and the number of dimensions along which it can occur is very large. In cybersecurity, hackers can adapt in unexpected ways, and thousands of extracted features are needed to categorize each software program or event. This differs from more conventional AI tasks, such as facial recognition, which have thousands of features to consider, but the information contained in the features does not shift over time. This means that the straightforward mitigation for distributional shift—expanding the initial dataset to incorporate potential future shifts—is simply infeasible. For example, one of the datasets uses 2,000 dimensions to categorize potential malware. Expanding the search space by just 10 percent along each dimension would increase the total volume of the search space by many orders of magnitude.

---

[16] Shu-Yi Xie, Jian Ma, Yu-Bin Luo, Lian-Xin Jiang, Shirly Jin, Yang Mo, and Jian-Ping Shen, "Models and Features with Covariate Shift Adaptation for Suspicious Network Event Recognition," *2019 IEEE International Conference on Big Data*, Los Angeles, December 9–12, 2019, p. 5946.

Approach

On the basis of our review of the literature, we selected two common cybersecurity tasks—network intrusion detection and malware identification—for which there are large, publicly available, benchmark datasets. For the network intrusion detection task, we built two AI models using the open-source framework TensorFlow to analyze two Canadian Institute for Cybersecurity (CIC) datasets: CIC-IDS2017 and CSE-CIC-IDS2018.[17] For the malware identification task, we used AI systems created specifically for three very large, interrelated datasets: Endgame Malware BEnchmark for Research (EMBER), Sophos-Reversing Labs–20 Million (SOREL-20M), and Blue Hexagon Open Dataset for Malware Analysis (BODMAS).[18]

We used these AI models to investigate how much distributional shift was present *within* these datasets and whether this shift was sufficient to degrade AI performance. To do this, we used a novel data segmentation process: We sliced the dataset into time windows and repeatedly trained and tested the *same* AI model on these different datasets. By freezing the AI architecture instead of optimizing the hyperparameters and other settings to each subset of data (which is the usual process), we could isolate the effects of changing data on AI performance. This enabled us to look for the presence of distributional shift and estimate the significance of its effect on AI performance over time.

Results

We confirmed that the cybersecurity datasets we studied suffer from distributional shift and that, to the extent that these datasets reflect reality, AI performance in these tasks should be expected to degrade over time. This means that AI systems for cybersecurity have an inherent shelf life and must be retrained regularly. There are two main findings:

- **Distributional shift is a significant factor for some cybersecurity applications, but timescales vary.** For network intrusion detection, both AI models exhibited a rapid degradation of about 4.5 percent in accuracy per day. For malware identification, we found a much slower change of about 1.5 percent in accuracy per year.[19] We emphasize, however, that

---

[17] Martín Abadi, Ashish Agarwal, Paul Barham, Eugene Brevdo, Zhifeng Chen, Craig Citro, Greg S. Corrado, Andy Davis, Jeffrey Dean, Matthieu Devin, Sanjay Ghemawat, Ian Goodfellow, Andrew Harp, Geoffrey Irving, Michael Isard, Rafal Jozefowicz, Yangqing Jia, Lukasz Kaiser, Manjunath Kudlur, Josh Levenberg, Dan Mané, Mike Schuster, Rajat Monga, Sherry Moore, Derek Murray, Chris Olah, Jonathon Shlens, Benoit Steiner, Ilya Sutskever, Kunal Talwar, Paul Tucker, Vincent Vanhoucke, Vijay Vasudevan, Fernanda Viégas, Oriol Vinyals, Pete Warden, Martin Wattenberg, Martin Wicke, Yuan Yu, and Xiaoqiang Zheng, *TensorFlow: Large-Scale Machine Learning on Heterogeneous Distributed Systems*, preliminary white paper, Google Research, November 9, 2015. The two CIC datasets are built in the same format for subsequent years. See Iman Sharafaldin, Arash Habibi Lashkari, and Ali A. Ghorbani, "Toward Generating a New Intrusion Detection Dataset and Intrusion Traffic Characterization," *Proceedings of the 4th International Conference on Information Systems Security and Privacy*, Funchal–Madeira, Portugal, January 22–24, 2018.

[18] Hyrum S. Anderson and Phil Roth, "EMBER: An Open Dataset for Training Static PE Malware Machine Learning Models," arXiv, April 16, 2018; Richard Harang and Ethan M. Rudd, "SOREL-20M: A Large Scale Benchmark Dataset for Malicious PE Detection," arXiv, December 14, 2020; Limin Yang, Arridhana Ciptadi, Ihar Laziuk, Ali Ahmadzadeh, and Gang Wang, "BODMAS: An Open Dataset for Learning Based Temporal Analysis of PE Malware," *2021 IEEE Symposium on Security and Privacy Workshops*, San Francisco, May 27, 2021.

[19] We used slightly different measures for each dataset because of their construction. For network intrusion detection, we measured accuracy using precision: true positives divided by all positives. For malware identification, we used the F_1 score, the harmonic mean of precision and recall.

these specific estimates arise from only a few periods of measurement, and other datasets might exhibit significantly different behavior.[20]

- **AI systems cannot reliably recognize new cyberattacks from old data.** Indeed, in the cases we considered, the AI systems could not recognize new attacks at all, but it is possible that more advanced AI models with more attuned hyperparameters could do a better job. However, when AI systems require recent data and data are only generated at a certain rate, such optimization can only go so far. This extends beyond cybersecurity to any AI algorithm facing unpredictable changes in a high-dimensional space.

From these findings, we obtain one general recommendation: **dataset segmentation tests like this can, and should, be performed to estimate the significance of distributional shift for any AI system for cybersecurity.** This kind of test yields a decay rate that provides an estimate of the retraining interval or shelf life for any AI system in this adaptive, complex space. It also indicates whether the AI system is generally sufficient or whether human oversight might be needed to address sudden new threats. This also requires AI developers to specify the training and testing datasets used—although this is widely recommended as a best practice, anyway.[21]

For more information on this use case, see Vol. 2, *Distributional Shift in Cybersecurity Datasets*.[22]

Predictive Maintenance

Our research objective for this use case was to consider how AI could improve failure analysis for aircraft parts, which is used to help determine the contents of readiness spares packages (RSPs).

USAF squadrons deploy with RSPs to support 30-day operations. The USAF currently uses the Aircraft Sustainability Model (ASM) to create a shopping list for each RSP on the basis of several factors, such as current supply, back order history, and estimates of part failure rates.[23] ASM uses a Poisson distribution to predict part failure rates.[24] In this research, we assessed the accuracy of part failure demand forecasting using a Poisson distribution and evaluated using AI as an alternative means to predict failure rates. It is important to note that ASM does much more than this: We consider only one of its many functions.

---

[20] In particular, while the network intrusion dataset represented real traffic running over a real computer network, it was created somewhat artificially on the basis of the actions of a small number of users over the course of a week. We do not know whether the threats represented were compressed in timescale or affected by the small number of users.

[21] Department of the Air Force and Massachusetts Institute of Technology, *Artificial Intelligence Acquisition Guidebook*, Artificial Intelligence Accelerator, February 14, 2022.

[22] Joshua Steier, Erik Van Hegewald, Anthony Jacques, Gavin S. Hartnett, and Lance Menthe, *Understanding the Limits of Artificial Intelligence for Warfighters: Vol. 2, Distributional Shift in Cybersecurity Datasets*, RAND Corporation, RR-A1722-2, 2024.

[23] F. Michael Slay, Tovey C. Bachman, Robert C. Kline, T. J. O'Malley, Frank L. Eichorn, and Randall M. King, *Optimizing Spares Support: The Aircraft Sustainability Model*, Logistics Management Institute, AF501MR1, October 1996.

[24] Other probability distributions, such as a negative binomial distribution, are technically available in the model, but they are not used by the USAF.

Approach

After considerable effort, we obtained a limited sample of historical data on ASM outputs and cross-referenced this sample with empirical data from the Logistics, Installations, and Mission Support–Enterprise View (LIMS-EV) system. This enabled us to compare predicted failure rates with actual failure rates to see how well the Poisson model used by ASM was actually working. We then built a proof-of-concept AI model to see whether it was possible to use AI to improve on these predictions.

Given the limitations of time and data, we looked only at part failures for a single platform, the A-10C, and used historical data pulled from September 2007 to June 2022. This time frame represented approximately 105,000 part failures for 113 unique part numbers.[25] We then built a long short-term memory model, a neural network designed to learn temporal relationships. We trained this AI model on the first ten years of data and tested on the data from the following years. Mindful of the potential for distributional shift, we then retrained the model on the basis of each month's data going forward, although we did not have time to perform the same kinds of data segmentation tests as we did for cybersecurity.

Results

A Kolmogorov-Smirnov test found that more than 80 percent of the failures did not match a Poisson distribution, which suggested that an AI model, or at least a non-Poisson distribution, had the potential to do better. Figure 2.1 offers a side-by-side comparison of the scatterplots showing how well the ASM model and the AI model were able to predict part failure rates.

---

[25] In the end, 16 percent of the data had to be discarded as ill-conditioned (e.g., lacking a date).

Figure 2.1. Comparison of Actual Part Failure Rates for the A-10C from 2007 to 2022 with Predictions by the Aircraft Sustainability Model and Artificial Intelligence Model

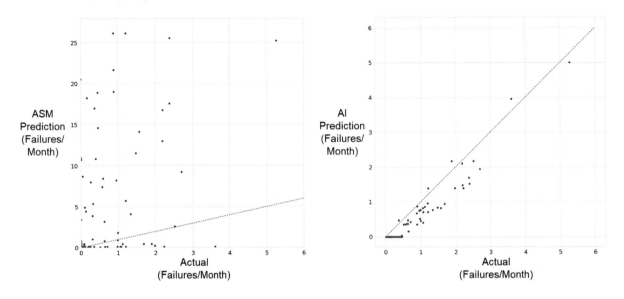

NOTE: The dotted red line indicates parity: If all predictions were accurate, they would all fall on the dotted red lines. These lines are tilted differently because the scales of the two charts differ: The scale of the ASM failure prediction rate axis is roughly four times higher than for the AI-predicted rates.

It is apparent to the eye that the predictions of the AI model are much closer than the Poisson distribution used by ASM. The ASM model has a wide scatter and significantly overstocks parts, while the AI predictions are almost always within one or two parts of the actual number. The average mean error of the AI model is 1.6 percent, resulting in $1.2 million in average monthly overstock costs; the average mean effort of the ASM model is 119 percent, resulting in $31 million in average monthly overstock costs.[26] Of course, there are good practical reasons why overstock might be desirable from a policy perspective, but likely not to this extent.

Despite these promising results, there are some important caveats. First, our AI model considers only parts for a single platform, whereas ASM must consider all platforms. The same ML approach used here can be reused to create new models for other platforms, although significant additional work might be needed to tune them, and we cannot know a priori if these models would show the same kinds of improvements as we found for the A-10C.

Second, as mentioned earlier, ASM does much more than predict part failures. It must also, for example, determine from which depots the parts will be pulled and to which bases they will be sent. This adds a complex operations research (OR) problem on top of the prediction problem. Significant additional work would be needed to address the more complicated question of selecting depots and bases. Furthermore, our work on mission planning, described in the "Mission Planning" section of this chapter, indicates that the application of AI to OR problems does not always improve optimality. (Although the benefits of the improved failure rate model would still hold.)

Finally, the RSP process involves forecasting rare events, which is inherently a difficult problem. A closer look at Figure 2.1 reveals that the AI model achieved some of its positive results simply by

---

[26] Although the AI model underpredicts failure, understock is not counted as a credit against overstock in this metric. Overstock and understock pose different policy challenges, and understocking generally is not beneficial.

predicting that some parts would never fail—a zero failure rate—which is clearly unsatisfactory for policy reasons. We built a separate proof-of-concept AI model specifically to examine these cases and achieved 77 percent accuracy to within plus or minus six months; however, while this is somewhat promising, the sample size was small. Likely, some hybrid would be needed for any real AI model. This is a particular issue for wartime failures, for which there is little available data. No matter how clever, AI cannot make up for a scarcity of data.

This proof of concept should not be viewed as a condemnation of ASM but rather as an invitation to experiment further in how the DAF might apply modern AI techniques to a venerable model.

From these results, we draw the following summary findings:

- **AI can improve demand forecasting for RSPs.** A static Poisson process is a poor predictor for many parts, and, likely, any single probability distribution would fall well short. We only showed this for a single platform and a single probability distribution, however, and we did not consider the larger issue of allocation from depots to bases.

- **A complex and labor-intensive data operations pipeline on DAF maintenance databases is necessary before any application of AI can occur.** Pulling LIMS-EV data is a manual process involving scripting, drop-down lists, and nested menus. It is practical only for a proof-of-concept model. Moreover, considerable data cleaning is necessary to unlock further historical data (e.g., linking variants prior to platform upgrades) and other potential predictors.

- **AI cannot alleviate the scarcity of wartime data.** Additional assumptions and policy considerations will be needed to account for this limitation. However, a regular retraining and updating policy, which is possible with an AI model, can assure adaptability when wartime comes.

From these findings, we make the following recommendations:

- **AFMC should build a data operations pipeline to conduct retrospective analysis of aircraft maintenance and RSP efficiency.** Aircraft maintenance programs and databases function effectively for the purposes for which they were designed, but they clearly were not designed for retrospective analysis or to train AI models. Unless the data can be properly conditioned and pulled for this analysis, none of the following recommendations can be implemented. As the *Artificial Intelligence Acquisition Guidebook* states, this is a common issue:

  > [J]ust gathering every piece of "data" does not solve our data asset crisis. Data can come in many forms, which may not be appropriate for the specific mission at hand. The PM [project manager], to the maximum extent possible, should request data in common formats and sizes to facilitate efficient data supply pipelines. In general, beneficial formats for AI projects include: image, video, tabular, comma-separated value (CSV), and/or tab-separated values (TSV). The proper data format will be dependent upon your AI algorithm and will be influenced by the type of model being trained. . . . Even with the proper file format, data needs to be curated and conditioned prior to AI training.[27]

---

[27] Department of the Air Force and Massachusetts Institute of Technology, 2022, p. 25.

- **Experiment with AI to improve demand forecasting for RSPs.** Extend the proof-of concept models to all aircraft. Although the modeling might need to be done on a part-by-part, platform-by-platform basis, doing so will be much simpler once the data operations pipeline is in place.
- **Consider using AI to solve the larger OR problem of selecting which part to send from which depot to which base.** This is a separate and complex problem.

For more information on this use case, see Vol. 3, *Predictive Maintenance*.[28]

Wargames

The research objective for this use case was to understand which aspects of wargames were most (and least) amenable to automation using AI systems.

In the 2010s, rapid progress in AI for playing games—Chess, Go, StarCraft II—inspired intense interest in the possible benefits of AI for playing wargames.[29] AI advocates suggested that AI might make wargames more effective or make it possible to apply wargames to novel kinds of problems. Recent RAND research, however, indicates that AI success at playing regular games does not carry over to real military C2 problems, so there is also reason for skepticism.[30] Moreover, there are many types of wargames and many parts of wargaming practice to which AI might apply, and it is reasonable to think that some are more promising for AI application than others.

Approach

The first step was to develop a general framework for thinking about how we design and execute wargames. We did so by reviewing the literature on AI in wargaming, both historical and contemporary, and interviewing SMEs, both wargame experts and AI practitioners. The framework we developed owes much to earlier RAND research by Elizabeth Bartels, who categorized wargames by purpose.[31] Table 2.1 summarizes these four types of wargames; the typical maturity of the tactics, techniques, and procedures (TTPs) employed; and the analytic focus.

---

[28] Li Ang Zhang, Yusuf Ashpari, and Anthony Jacques, *Understanding the Limits of Artificial Intelligence for Warfighters: Vol. 3, Predictive Maintenance*, RR-A1722-3, 2024.

[29] For example, see David Silver, David, Aja Huang, Chris J. Maddison, Arthur Guez, Laurent Sifre, George van den Driessche, Julian Schrittwieser, Ioannis Antonoglou, Veda Panneershelvam, Marc Lanctot, Sander Dieleman, Dominik Grewe, John Nham, Nal Kalchbrenner, Ilya Sutskever, Timothy Lillicrap, Madeleine Leach, Koray Kavukcuoglu, Thore Graepel, and Demis Hassabis, "Mastering the Game of Go with Deep Neural Networks and Tree Search," *Nature*, Vol. 529, January 27, 2016; and Oriol Vinyals, Igor Babuschkin, Wojciech M. Czarnecki, Michaël Mathieu, Andrew Dudzik, Junyoung Chung, David H. Choi, Richard Powell, Timo Ewalds, Petko Georgiev, Junhyuk Oh, Dan Horgan, Manuel Kroiss, Ivo Danihelka, Aja Huang, Laurent Sifre, Trevor Cai, John P. Agapiou, Max Jaderberg, Alexander S. Vezhnevets, Rémi Leblond, Tobias Pohlen, Valentin Dalibard, David Budden, Yury Sulsky, James Molloy, Tom L. Paine, Caglar Gulcehre, Ziyu Wang, Tobias Pfaff, Yuhuai Wu, Roman Ring, Dani Yogatama, Dario Wünsch, Katrina McKinney, Oliver Smith, Tom Schaul, Timothy Lillicrap, Koray Kavukcuoglu, Demis Hassabis, Chris Apps, and David Silver, "Grandmaster Level in StarCraft II Using Multi-Agent Reinforcement Learning," *Nature*, Vol. 575, October 30, 2019.

[30] Walsh et al., 2021a.

[31] Elizabeth M. Bartels, *Building Better Games for National Security Policy Analysis: Towards a Social Scientific Approach*, dissertation, Pardee RAND Graduate School, RAND Corporation, RGSD-437, 2020.

Table 2.1. Taxonomy of Wargames by Purpose

Type	Purpose	TTPs	Analytic Focus	Example
Systems exploration	Improve understanding of tactical or technological problems by eliciting and synthesizing mental models of the problem from expert players.	Evolving	Problem	RAND's Gray Zone Games[a]
Innovation	Propose new decision options outside the status quo, to "think outside the box" in terms of solutions.	Evolving	Solution	DARPA's Mosaic Warfare Games[b]
Alternative conditions	Examine how decisionmaking does or does not change in response to variations in starting conditions, to test strategic robustness.	Mature	Problem	USAF Strategy and Force Evaluation (SAFE) Games[c]
Evaluation	Judge the outcomes of player decisions against some normative standard.	Mature	Solution	DARPA/CNA "Scud Hunt"[d]

SOURCE: Adapted from Bartels, 2020.
NOTE: CNA = Center for Naval Analyses; DARPA = Defense Advanced Research Projects Agency.
[a] Becca Wasser, Jenny Oberholtzer, Stacie L. Pettyjohn, and William Mackenzie, *Gaming Gray Zone Tactics: Design Considerations for a Structured Strategic Game*, RAND Corporation, RR-2915-A, 2019.
[b] Timothy R. Gulden, Jonathan Lamb, Jeff Hagen, and Nicholas A. O'Donoughue, *Modeling Rapidly Composable, Heterogeneous, and Fractionated Forces: Findings on Mosaic Warfare from an Agent-Based Model*, RAND Corporation, RR-4396-OSD, 2021.
[c] Thomas A. Brown and Edwin W. Paxson, *A Retrospective Look at Some Strategy and Force Evaluation Games*, RAND Corporation, R-1619-PR, 1975.
[d] Peter P. Perla, Michael Markowitz, Albert Nofi, Christopher Weuve, Julia Loughran, and Marcy Stahl, *Gaming and Shared Situation Awareness*, Center for Naval Analyses, November 2000.

Although these types of games are not mutually exclusive, they call for different approaches to planning and execution that are in tension with one another. For example, the rigidity of form that makes alternative conditions and evaluation games useful likely would be counterproductive for systems evaluation and innovation.

We extended this framework to separate the time-phased tasks of wargames: *preparing, playing, adjudicating,* and *interpreting.* This grouping by purpose and division by task was the general framework for analysis. We then estimated the technical feasibility and cost-effectiveness of applying AI to each task and purpose. This was primarily done through discussions with SMEs but also following theoretical considerations from which no previous attempt to apply AI to this area could be identified.

Results

Figure 2.2 shows our estimate of the combined technical feasibility and cost-effectiveness of developing and deploying AI for wargames by type (purpose) and time-phased task.

Figure 2.2. Technical Feasibility and Cost-Effectiveness of Developing Artificial Intelligence for Wargames

Task / Type	Preparing	Playing	Adjudicating	Interpreting
Systems Exploration	Feasible and affordable when the problem specification is simple and available, especially for repeated games	Feasible for well-defined domains	Feasible when adjudication terms are straightforward	Requires human-level AI
Innovation		Requires human-level AI	Possibility of unforeseen player actions complicates adjudication	Requires human-level AI
Alternative Conditions		Feasible for well-defined domains	Feasible when adjudication terms are straightforward	Feasible except where differences are very complex
Evaluation		Feasible if low-quality players are adequate for game application	Feasible when adjudication terms are straightforward	Feasible except where goals are not well specified

Prohibitive	Possible	Feasible

The figure is presented as a traditional stoplight chart: Red is prohibitively expensive or difficult, green is relatively easy or affordable, and yellow is somewhere in between. We also have intermediate rankings for a few areas. It is immediately evident that not much is easy and there are many areas where AI is very limited. Nevertheless, there remain opportunities for investing in AI for wargames today, and there are other areas where appropriate investments in technology could move the rating from yellow to green.

We find that AI application is most promising for wargames that

- are designed to investigate alternative conditions or are used for evaluation, especially those with a small set of well-defined evaluation criteria
- employ computational models in a significant role during the adjudication process or that generate large volumes of digital data that must be adjudicated
- use advanced human-computer interaction (HCI) technologies for data capture and interaction (e.g., cameras and microphones)
- are repeated many times, especially those that model zero-sum, force-on-force conflicts.

In contrast, AI application appears to be least promising for wargames that

- are designed for systems exploration or innovation
- employ limited digital infrastructure or interaction with computational models
- are conducted in security environments in which advanced HCI technologies are restricted
- are played as one-offs or a very limited number of times for specific purposes.

From these determinations, we make three recommendations for how to direct investment in AI to support future wargames.

1. **Resources devoted to developing AI applications for wargames should be concentrated on the most promising areas**: those that investigate alternative conditions or are used for

evaluation, with well-defined problems and criteria; those that already incorporate digital infrastructure, including HCI technologies; and those that are regularly repeated.

2. **The use of digital gaming infrastructure and HCI technologies should be increased**, especially in games designed for systems exploration and innovation. The digitization of wargaming tasks must precede the future application of AI to many wargaming tasks. HCI technologies can and should be employed to gather data on discourse and decisionmaking to support AI development. This could be of particular value in cases in which AI can learn from evaluation-type wargames.

3. **AI capabilities should be employed in strategic studies to support future wargaming efforts more generally**, to shift items from the possible to the feasible. These studies include scenario generation and case identification to find challenging conditions that merit the attention of games and sentiment or stance analysis in support of qualitative research on wargames.

For more information on this use case, see Vol. 4, *Wargames*.[32]

Mission Planning

The research objective for this use case was to find the limits of when and how AI methods can improve on more conventional OR methods for mission planning (specifically, route planning).

The general problem of mission planning involves the simultaneous assignment of multiple assets to prioritized targets, including dynamic routing to their destinations under complex environmental conditions. Previous RAND research on the deliberate Master Air Attack Plan process identified low data availability and operational risks as the most challenging constraints to AI implementation for this case.[33] An experiment conducted during that research showed that a mixed integer program, a more conventional OR method, was able to achieve very close to the optimal solution but took several hours to do so, whereas a simple ML method could come "within 17 percent of optimality after two seconds."[34] Moreover, the case considered was relatively simple: The explosion of possible combinations that would result from considering thousands of assets would render conventional methods obsolete. Meanwhile, a more detailed look at AI for mission planning in the Advanced Framework for Simulation, Integration, and Modeling (AFSIM) showed some promise in this area, but only one of the many algorithms tested was able to generate reasonable routes reliably.[35]

In this research, we built on this previous modeling and simulation work with AFSIM to understand under which conditions standard AI methods could equal or exceed OR methods.

---

[32] Edward Geist, Aaron B. Frank, and Lance Menthe, *Understanding the Limits of Artificial Intelligence for Warfighters: Vol. 4, Wargames*, RR-A1722-4, 2024.

[33] Walsh et al., 2021a.

[34] Matthew Walsh, Lance Menthe, Edward Geist, Eric Hastings, Joshua Kerrigan, Jasmin Léveillé, Joshua Margolis, Nicholas Martin, and Brian P. Donnelly, *Exploring the Feasibility and Utility of Machine Learning–Assisted Command and Control: Vol. 2, Supporting Technical Analysis*, RAND Corporation, RR-A263-2, 2021b, p. 56.

[35] Li Ang Zhang, Jia Xu, Dara Gold, Jeff Hagen, Ajay K. Kochhar, Andrew J. Lohn, and Osonde A. Osoba, *Air Dominance Through Machine Learning: A Preliminary Exploration of Artificial Intelligence–Assisted Mission Planning*, RAND Corporation, RR-4311-RC, 2020.

15

Approach

We used the RAND Target Acquisition Model (RTAM) of one drone attempting to avoid detection as it navigates to a point. We compared two navigation methods. One was a more conventional method that divided the airspace into discrete spaces and selected the best route between these spaces on the basis of an assigned cost function. Because of the relatively small number of discrete spaces, it is possible to find the mathematically optimal space given those constraints in a reasonable time frame, even though the problem grows in complexity in polynomial time. The other was an AI method based on reinforcement learning in AFSIM that sought to discover the best path between these spaces in an open-ended manner. Screenshots of both models are shown in Figure 2.3.

Figure 2.3. Screenshots of the RAND Target Acquisition Model and Advanced Framework for Simulation, Integration, and Modeling

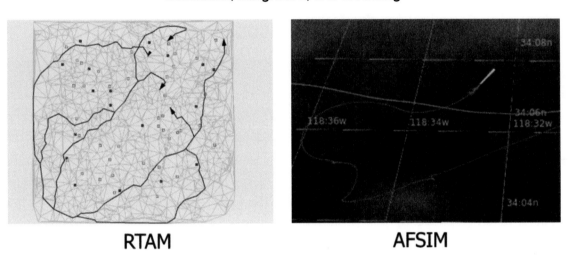

RTAM   AFSIM

As part of this research, we created a pipeline between AFSIM and a widely used ML framework, which has been shared with AFMC.[36] We produced an end-to-end integration that enabled the training of an AI model in AFSIM to perform basic tasks. Importantly, this version does not require a Python harness, which would cause substantial delays to runtime.[37] Trained AI models instead can be deployed directly to AFSIM.[38] We used this to implement AI in AFSIM to test when it would be more efficient than other methods.

Results

In our results, OR methods are always more optimal than AI methods. This might seem surprising, but in this case, the OR method yields the mathematically maximal value, while AI

---

[36] Gary J. Briggs, *AFSIM Reinforcement Learning Tool*, RAND Corporation, TL-A1722-1, 2022, Not available to the general public.

[37] Many AI models are written in Python, so a special interface is needed to use these models in other programming environments.

[38] Because AFSIM supports C++ commands and TensorFlowLite exports to C++, it is possible to run an AI model without involving any Python.

methods are inherently function approximators. However, AI has three distinct advantages. First, the solutions it generates are often more robust, meaning that the paths might be suboptimal, but they remain viable if the threats move; second, it can consider dynamic pop-up targets without retraining; and third, the solution is often developed much faster.

From these results, we extract the following major findings:

- AI-based mission planning produces less optimal results than traditional OR methods for well-defined route-finding problems.
- AI produces many paths that are near-optimal, produces more-robust solutions, and can develop them much faster.
- AI can augment mission planning by generating potential paths for human consideration because it considers many options explicitly.
- Current technological limitations prevent AI from being applied to long-term strategic-level planning.

From these findings we draw the following recommendations:

- **The DAF should invest in developing appropriate tools to enable reinforcement learning models to be applied to existing mission planning models.** Applying such approaches within AFSIM can assist in tactical mission planning as well.
- **The DAF should consider how AI could power a fast-reaction policy for drones facing unexpected conditions.** We demonstrated efficient small models that can run onboard small drones.

For more information on this use case, see Vol. 5, *Mission Planning*.[39]

Conclusion

Across all use cases, two common themes emerged: (1) Data to train and test AI systems must be current, accessible, available, and of high quality, and the scarcity of such data significantly restricts the effectiveness of AI for warfighting applications, especially where this scarcity cannot be mitigated; and (2) the limitations of AI algorithms, in terms of how and what they learn, can significantly restrict their utility to specialized uses, especially where human insights are involved. *AI clearly has potential to benefit all four use cases*, but because of these limits, its application suffers from significant constraints on its use. Table 2.2 summarizes the major findings for each use case, grouped by these common themes.

---

[39] Keller Scholl, Gary J. Briggs, Li Ang Zhang, and John L. Salmon, *Understanding the Limits of Artificial Intelligence for Warfighters: Vol. 5, Mission Planning*, RR-A1722-5, 2024.

Table 2.2. Summary of Findings

Use Case	Data Limitations	Algorithm Limitations
Cybersecurity	• **To recognize adaptive threats, data must be recent.** Distributional shift degrades model performance, and it cannot be avoided, especially for high-dimensional data.	• **AI classification algorithms cannot be relied on to learn what they are not taught.** AI did not anticipate or recognize new kinds of cyberattacks.
Predictive maintenance	• **Data must be accessible and well-conditioned.** Relevant logistics data are maintained in multiple databases and are often ill-conditioned. Without an automated data pipeline, sufficient data cannot be captured to enable AI. • **Peacetime data cannot be substituted for wartime data.** AI cannot make up for a scarcity of appropriate data.	• **AI can estimate complex functions well, but this comes at a loss of generality.** AI can estimate part failure rates far better than a universal probability distribution, but it must be trained on each part separately.
Wargames	• **Digitization must precede AI development.** Most wargames are not conducted in a digital environment and do not generate electronic data. Digitization is a precursor to an AI data pipeline. • **New kinds of data are needed.** To enable AI, HCI technology is needed to capture aspects of wargaming that are not captured today.	• **AI is far from achieving human-level intelligence.** Therefore, it cannot stand in for humans, nor can it apply human judgments. AI is therefore only likely applicable to certain stages of wargames conducted for certain purposes.
Mission planning	• **To counter adaptive threats, data must be recent.** Models must be refreshed with updated conditions to survive against dynamic threats.	• **AI is tactically brilliant but strategically naive.** It tends to win by getting within the opponent's observe, orient, decide, act loop rather than by coming up with a clever grand strategy. • **AI is less accurate than traditional optimization methods.** But its solutions can be more robust, and it can reach them faster.

In the recommendations across all use cases, two themes emerged: the need to conduct tests and experiments and the need to develop better infrastructure to support future AI development, including digitization technologies and data pipelines. Table 2.3 summarizes the recommendations, specific to each use case.

Table 2.3. Summary of Recommendations

Use Case	AI Tests and Experiments	Supporting Infrastructure
Cybersecurity	• Perform dataset segmentation tests to determine the significance of distributional shift for AI systems and to determine an approximate decay rate and AI shelf life.	• Not applicable.
Predictive maintenance	• Experiment with AI to improve demand forecasting for RSPs and extend the proof-of-concept models to all aircraft. This will likely have to be done on a part-by-part, platform-by-platform basis. • Consider AI to solve the larger OR problem of selecting which parts to send where.	• Build a data operations pipeline to conduct a retrospective analysis of aircraft maintenance and RSP efficiently for multiple parts and platforms.
Wargames	• Concentrate resources for developing AI applications for wargames on the most promising areas: those that investigate alternative conditions or that are used for evaluation with well-defined criteria; those that already incorporate digital infrastructure, including HCI technologies; and those that are regularly repeated.	• Increase the use of digital gaming infrastructure and HCI technologies, especially in games designed for systems exploration and innovation, to gather data to support AI development. • Employ AI capabilities to support future wargaming efforts more generally.
Mission planning	• Consider how AI could power a fast-reaction policy for drones facing unexpected conditions.	• Invest in developing tools to apply reinforcement learning to existing mission planning models and in simulations, such as AFSIM.

There is much work to be done to answer the overall question of the limits of AI for warfighting applications. We have found several limits, some general, some specific, and no doubt there are many others. AI is not a panacea. There is much it can do but also much that it cannot. This is a promising area for tests and experiments, but data pipelines and other technologies are necessary to enable that research. The DAF should support both.

Abbreviations

AFMC	Air Force Materiel Command
AFSIM	Advanced Framework for Simulation, Integration, and Modeling
AI	artificial intelligence
ASM	Aircraft Sustainability Model
C2	command and control
CIC	Canadian Institute for Cybersecurity
DAF	Department of the Air Force
HCI	human-computer interaction
ISR	intelligence, surveillance, and reconnaissance
LIMS-EV	Logistics, Installations and Mission Support–Enterprise View
ML	machine learning
OR	operations research
RSP	readiness spares package
RTAM	RAND Target Acquisition Model
SME	subject-matter expert
TTPs	tactics, techniques, and procedures
USAF	U.S. Air Force

References

Abadi, Martín, Ashish Agarwal, Paul Barham, Eugene Brevdo, Zhifeng Chen, Craig Citro, Greg S. Corrado, Andy Davis, Jeffrey Dean, Matthieu Devin, Sanjay Ghemawat, Ian Goodfellow, Andrew Harp, Geoffrey Irving, Michael Isard, Rafal Jozefowicz, Yangqing Jia, Lukasz Kaiser, Manjunath Kudlur, Josh Levenberg, Dan Mané, Mike Schuster, Rajat Monga, Sherry Moore, Derek Murray, Chris Olah, Jonathon Shlens, Benoit Steiner, Ilya Sutskever, Kunal Talwar, Paul Tucker, Vincent Vanhoucke, Vijay Vasudevan, Fernanda Viégas, Oriol Vinyals, Pete Warden, Martin Wattenberg, Martin Wicke, Yuan Yu, and Xiaoqiang Zheng, *TensorFlow: Large-Scale Machine Learning on Heterogeneous Distributed Systems*, preliminary white paper, Google Research, November 9, 2015.

Anderson, Hyrum S., and Phil Roth, "EMBER: An Open Dataset for Training Static PE Malware Machine Learning Models," arXiv, April 16, 2018.

Arai, Kohei, and Supriya Kapoor, eds, *Advances in Computer Vision: Proceedings of the 2019 Computer Vision Conference CVC*, Vol. 1, Springer, 2019.

Austin, Lloyd J., III, "Secretary of Defense Austin Remarks at the Global Emerging Technology Summit of the National Security Commission on Artificial Intelligence (as Delivered)," transcript, U.S. Department of Defense, July 13, 2021.

Bartels, Elizabeth M., *Building Better Games for National Security Policy Analysis: Towards a Social Scientific Approach*, dissertation, Pardee RAND Graduate School, RAND Corporation, RGSD-437, 2020. As of August 27, 2022:
https://www.rand.org/pubs/rgs_dissertations/RGSD437.html

Bishop, Christopher M., *Pattern Recognition and Machine Learning*, Springer, 2006.

Briggs, Gary J., *AFSIM Reinforcement Learning Tool*, RAND Corporation, TL-A1722-1, 2022, Not available to the general public.

Brown, Thomas A., and Edwin W. Paxson, *A Retrospective Look at Some Strategy and Force Evaluation Games*, RAND Corporation, R-1619-PR, 1975. As of September 13, 2023:
https://www.rand.org/pubs/reports/R1619.html

Department of the Air Force and Massachusetts Institute of Technology, *Artificial Intelligence Acquisition Guidebook*, Artificial Intelligence Accelerator, February 14, 2022.

Egel, Daniel, Ryan Andrew Brown, Linda Robinson, Mary Kate Adgie, Jasmin Léveillé, and Luke J. Matthews, *Leveraging Machine Learning for Operation Assessment*, RAND Corporation, RR-4196-A, 2022. As of August 25, 2022:
https://www.rand.org/pubs/research_reports/RR4196.html

Geist, Edward, Aaron B. Frank, and Lance Menthe, *Understanding the Limits of Artificial Intelligence for Warfighters: Vol. 4, Wargames*, RAND Corporation, RR-A1722-4, 2024.

Gulden, Timothy R., Jonathan Lamb, Jeff Hagen, and Nicholas A. O'Donoughue, *Modeling Rapidly Composable, Heterogeneous, and Fractionated Forces: Findings on Mosaic Warfare from an Agent-Based Model*, RAND Corporation, RR-4396-OSD, 2021. As of September 13, 2023:
https://www.rand.org/pubs/research_reports/RR4396.html

Harang, Richard, and Ethan M. Rudd, "SOREL-20M: A Large Scale Benchmark Dataset for Malicious PE Detection," arXiv, December 14, 2020.

Ish, Daniel, Jared Ettinger, and Christopher Ferris, *Evaluating the Effectiveness of Artificial Intelligence Systems in Intelligence Analysis*, RAND Corporation, RR-A464-1, 2021. As of August 25, 2022: https://www.rand.org/pubs/research_reports/RRA464-1.html

Jordan, M. I., and T. M. Mitchell, "Machine Learning: Trends, Perspectives, and Prospects," *Science*, Vol. 349, No. 6245, July 2015.

Lingel, Sherrill, Jeff Hagen, Eric Hastings, Mary Lee, Matthew Sargent, Matthew Walsh, Li Ang Zhang, and David Blancett, *Joint All-Domain Command and Control for Modern Warfare: An Analytic Framework for Identifying and Developing Artificial Intelligence Applications*, RAND Corporation, RR-4408/1-AF, 2020. As of August 25, 2022: https://www.rand.org/pubs/research_reports/RR4408z1.html

Lohn, Andrew J., Jair Aguirre, Mark Ashby, Benjamin Boudreaux, Johnathan Fujiwara, Gavin S. Hartnett, Daniel Ish, John Speed Meyers, Caolionn O'Connell, and Li Ang Zhang, *Attacking Machine Learning in War*, RAND Corporation, RR-4386-AF, 2020, Not available to the general public.

Marcus, Gary, "Artificial General Intelligence Is Not as Imminent as You Might Think," *Scientific American*, July 1, 2022.

Menthe, Lance, Dahlia Anne Goldfeld, Abbie Tingstad, Sherrill Lingel, Edward Geist, Donald Brunk, Amanda Wicker, Sarah Lovell, Balys Gintautas, Anne Stickells, and Amado Cordova, *Technology Innovation and the Future of Air Force Intelligence Analysis:* Vol. 1, *Findings and Recommendations*, RAND Corporation, RR-A341-1, 2021a. As of August 25, 2022: https://www.rand.org/pubs/research_reports/RRA341-1.html

Menthe, Lance, Dahlia Anne Goldfeld, Abbie Tingstad, Sherrill Lingel, Edward Geist, Donald Brunk, Amanda Wicker, Sarah Lovell, Balys Gintautas, Anne Stickells, and Amado Cordova, *Technology Innovation and the Future of Air Force Intelligence Analysis:* Vol. 2, *Technical Analysis and Supporting Material*, RAND Corporation, RR-A341-2, 2021b. As of April 13, 2022: https://www.rand.org/pubs/research_reports/RRA341-2.html

Minsky, Marvin, ed., *Semantic Information Processing*, MIT Press, 1968.

O'Mahony, Niall, Sean Campbell, Anderson Carvalho, Suman Harapanahalli, Gustavo Velasco Hernández, Lenka Krpalkova, Daniel Riordan, and Joseph Walsh, "Deep Learning vs. Traditional Computer Vision," in Kohei Arai and Supriya Kapoor, eds, *Advances in Computer Vision: Proceedings of the 2019 Computer Vision Conference CVC*, Vol. 1, Springer, 2019.

Perla, Peter P., Michael Markowitz, Albert Nofi, Christopher Weuve, Julia Loughran, and Marcy Stahl, *Gaming and Shared Situation Awareness*, Center for Naval Analyses, November 2000.

Robson, Sean, Maria C. Lytell, Matthew Walsh, Kimberly Curry Hall, Kirsten M. Keller, Vikram Kilambi, Joshua Snoke, Jonathan W. Welburn, Patrick S. Roberts, Owen Hall, and Louis T. Mariano, *U.S. Air Force Enlisted Classification and Reclassification: Potential Improvements Using Machine Learning and Optimization Models*, RAND Corporation, RR-A284-1, 2022. As of August 25, 2022: https://www.rand.org/pubs/research_reports/RRA284-1.html

Russell, Stuart, and Peter Norvig, *Artificial Intelligence: A Modern Approach*, 4th ed., Pearson, 2021.

Scholl, Keller, Gary J. Briggs, Li Ang Zhang, and John L. Salmon, *Understanding the Limits of Artificial Intelligence for Warfighters:* Vol. 5, *Mission Planning*, RAND Corporation, RR-A1722-5, 2024.

Schulker, David, Nelson Lim, Luke J. Matthews, Geoffrey E. Grimm, Anthony Lawrence, and Perry Shameem Firoz, *Can Artificial Intelligence Help Improve Air Force Talent Management? An Exploratory Application*, RAND Corporation, RR-A812-1, 2021. As of August 25, 2022: https://www.rand.org/pubs/research_reports/RRA812-1.html

Sharafaldin, Iman, Arash Habibi Lashkari, and Ali A. Ghorbani, "Toward Generating a New Intrusion Detection Dataset and Intrusion Traffic Characterization," *Proceedings of the 4th International Conference on Information Systems Security and Privacy*, Funchal–Madeira, Portugal, January 22–24, 2018.

Silver, David, Aja Huang, Chris J. Maddison, Arthur Guez, Laurent Sifre, George van den Driessche, Julian Schrittwieser, Ioannis Antonoglou, Veda Panneershelvam, Marc Lanctot, Sander Dieleman, Dominik Grewe, John Nham, Nal Kalchbrenner, Ilya Sutskever, Timothy Lillicrap, Madeleine Leach, Koray Kavukcuoglu, Thore Graepel, and Demis Hassabis, "Mastering the Game of Go with Deep Neural Networks and Tree Search," *Nature*, Vol. 529, January 27, 2016.

Slay, F. Michael, Tovey C. Bachman, Robert C. Kline, T. J. O'Malley, Frank L. Eichorn, and Randall M. King, *Optimizing Spares Support: The Aircraft Sustainability Model*, Logistics Management Institute, AF501MR1, October 1996.

Steier, Joshua, Erik Van Hegewald, Anthony Jacques, Gavin S. Hartnett, and Lance Menthe, *Understanding the Limits of Artificial Intelligence for Warfighters: Vol. 2, Distributional Shift in Cybersecurity Datasets*, RAND Corporation, RR-A1722-2, 2024.

Vinyals, Oriol, Igor Babuschkin, Wojciech M. Czarnecki, Michaël Mathieu, Andrew Dudzik, Junyoung Chung, David H. Choi, Richard Powell, Timo Ewalds, Petko Georgiev, Junhyuk Oh, Dan Horgan, Manuel Kroiss, Ivo Danihelka, Aja Huang, Laurent Sifre, Trevor Cai, John P. Agapiou, Max Jaderberg, Alexander S. Vezhnevets, Rémi Leblond, Tobias Pohlen, Valentin Dalibard, David Budden, Yury Sulsky, James Molloy, Tom L. Paine, Caglar Gulcehre, Ziyu Wang, Tobias Pfaff, Yuhuai Wu, Roman Ring, Dani Yogatama, Dario Wünsch, Katrina McKinney, Oliver Smith, Tom Schaul, Timothy Lillicrap, Koray Kavukcuoglu, Demis Hassabis, Chris Apps, and David Silver, "Grandmaster Level in StarCraft II Using Multi-Agent Reinforcement Learning," *Nature*, Vol. 575, October 30, 2019.

Walsh, Matthew, Lance Menthe, Edward Geist, Eric Hastings, Joshua Kerrigan, Jasmin Léveillé, Joshua Margolis, Nicholas Martin, and Brian P. Donnelly, *Exploring the Feasibility and Utility of Machine Learning-Assisted Command and Control: Vol. 1, Findings and Recommendations*, RAND Corporation, RR-A263-1, 2021a. As of August 25, 2022: https://www.rand.org/pubs/research_reports/RRA263-1.html

Walsh, Matthew, Lance Menthe, Edward Geist, Eric Hastings, Joshua Kerrigan, Jasmin Léveillé, Joshua Margolis, Nicholas Martin, and Brian P. Donnelly, *Exploring the Feasibility and Utility of Machine Learning-Assisted Command and Control: Vol. 2, Supporting Technical Analysis*, RAND Corporation, RR-A263-2, 2021b. As of September 5, 2022: https://www.rand.org/pubs/research_reports/RRA263-2.html

Wasser, Becca, Jenny Oberholtzer, Stacie L. Pettyjohn, and William Mackenzie, *Gaming Gray Zone Tactics: Design Considerations for a Structured Strategic Game*, RAND Corporation, RR-2915-A, 2019. As of September 14, 2023: https://www.rand.org/pubs/research_reports/RR2915.html

Xie, Shu-Yi, Jian Ma, Yu-Bin Luo, Lian-Xin Jiang, Shirly Jin, Yang Mo, and Jian-Ping Shen, "Models and Features with Covariate Shift Adaptation for Suspicious Network Event Recognition," *2019 IEEE International Conference on Big Data*, Los Angeles, December 9–12, 2019.

Yang, Limin, Arridhana Ciptadi, Ihar Laziuk, Ali Ahmadzadeh, and Gang Wang, "BODMAS: An Open Dataset for Learning Based Temporal Analysis of PE Malware," *2021 IEEE Symposium on Security and Privacy Workshops*, San Francisco, May 27, 2021.

Zhang, Li Ang, Yusuf Ashpari, and Anthony Jacques, *Understanding the Limits of Artificial Intelligence for Warfighters: Vol. 3, Predictive Maintenance*, RAND Corporation, RR-A1722-3, 2024.

Zhang, Li Ang, Gavin S. Hartnett, Jair Aguirre, Andrew J. Lohn, Inez Khan, Marissa Herron, and Caolionn O'Connell, *Operational Feasibility of Adversarial Attacks Against Artificial Intelligence*, RAND Corporation, RR-A866-1, 2022. As of September 14, 2023:
https://www.rand.org/pubs/research_reports/RRA866-1.html

Zhang, Li Ang, Lance Menthe, Ian Fleischmann, Sale Lilly, Joshua Kerrigan, Michael J. Gaines, and Gregory A. Schumacher, *Incorporating Artificial Intelligence into Army Intelligence Processes*, RAND Corporation, RR-A729-1, 2021, Not available to the general public.

Zhang, Li Ang, Jia Xu, Dara Gold, Jeff Hagen, Ajay K. Kochhar, Andrew J. Lohn, and Osonde A. Osoba, *Air Dominance Through Machine Learning: A Preliminary Exploration of Artificial Intelligence–Assisted Mission Planning*, RAND Corporation, RR-4311-RC, 2020. As of August 25, 2022:
https://www.rand.org/pubs/research_reports/RR4311.html